THE MYSTERIOUS LINDSEY LIGHTS

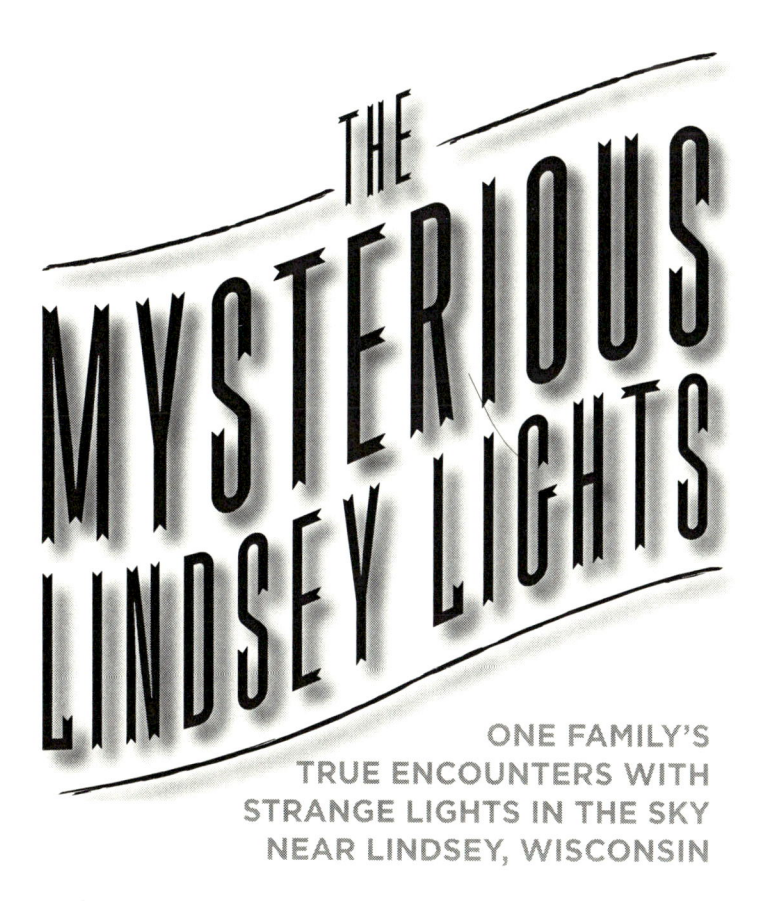

ONE FAMILY'S TRUE ENCOUNTERS WITH STRANGE LIGHTS IN THE SKY NEAR LINDSEY, WISCONSIN

Betty May Ockerl

VANTAGE Press
NEW YORK

Illustrations by Betty May Ockerl
Cover design by Neuwirth & Associates
Vantage Press and the Vantage Press colophon
are registered trademarks of Vantage Press, Inc.

FIRST EDITION
All rights reserved, including the right of
reproduction in whole or in part in any form.
Copyright © 2011 by Betty May Ockerl
Published by Vantage Press, Inc.

419 Park Ave. South, New York, NY 10016

Manufactured in the United States of America
ISBN: 978-0-533163-99-1

Library of Congress Catalog Card No: 2010909903

0 9 8 7 6 5 4 3 2 1

Dedication

I dedicate this book to my daughter, Amanda, and to those individuals who challenge the thought that we may not be alone. I also dedicate it to "those" who may have already visited here.

CONTENTS

Foreword vii

Acknowledgements xi

Introduction xiii

1 Do You See Those Lights? 1

2 Some Strange 'Moon'! 7

3 Read All About It 13

4 They're Back! 17

5 When You Least Expect It 21

6 'Sparkles' and Shine 25

7 Big Red! 29

8 One Starry Night 35

9   Jet Action   39

10   We're Surrounded!   45

11   That Was No Plane!   51

12   Mysterious Lights   59

13   They're Still Out There!   63

14   The Believers   69

15   It Was Only a Dream   73

16   Doctor, Doctor   77

Conclusion   83

FOREWORD

Back in the late 1960's at a summer camp in Wisconsin, a girlfriend and I went for a night walk through the woods to a clearing behind the camp area. We stopped on a hill and sat down to look at the sky and talk about the stars.

Suddenly, we saw what appeared to be a bright light crossing the sky. It seemed to stop and begin to come straight down. It did a complete circle and hovered there like it was studying the area and preparing to land. There was no sound or wind. We were fairly close to this thing, and to us, it seemed very odd. We got so scared we ran back to the camp.

The next day, we went back to look for it, and all we saw was the clearing. There were no roads close by so it couldn't have been a vehicle. And because it had no flashing lights, it didn't resemble any aircraft. We knew what we saw could have been a UFO!

One day in 2005, Betty phoned me because she saw my name while reading the *Marshfield News-Herald*. The article she was reading was about the paranormal, and I had mentioned that I had seen some strange lights. We had a long discussion about numerous different sightings we had both seen. Since then, Betty has become a good friend of mine.

Through our discussions, and a little detective work, we both found out that there were more people from the Lindsey area that had seen possible UFOs. That is why Betty decided to take the challenge and write her own book. I want to note her bravery for writing about a subject that most people would find uneasy to talk about. Betty's book is not to prove the existence of UFOs, rather, it is to give you a real eyewitness account of something truly strange going on.

One thing is for sure, the universe is a huge,

unexplored place, and because of that fact, the subject of UFOs should not be taken for granted. I personally think there is something to UFO sightings, and we will some day know the truth!

I guarantee Betty's book, *The Mysterious Lindsey Lights*, will be a real adventure to read. Maybe then you will also watch the sky and witness for yourself the lights of our visiting friends.

I would like to thank Betty for letting me be one of her witnesses and for giving me the opportunity to write the foreword for her book.

Diane Luchterhand
July, 2006

ACKNOWLEDGEMENTS

I would like to thank all my friends and witnesses for coming forth about their own UFO sightings. Although some of their encounters are not in my book, their stories gave me great encouragement. It was good to know that our family was not the only one to see strange lights around the Lindsey area. Thank you all.

INTRODUCTION

When Amanda was born in 1993, we were renting a farmhouse about fifteen miles west of the city of Marshfield. Gene and I both grew up on farms and we planned on living in the country for the rest of our lives.

One day, we decided we wanted to buy our own house on at least five acres of land, so I placed an ad in the newspaper and started looking for a house in the country. After looking at several places close to neighboring towns, we grew discouraged. Then, finally, we received a call from a young couple selling their small house with five acres of woods. It sounded perfect! We made the sellers an offer on an additional five acres of woods. They accepted our offer. We couldn't have been more excited!

We quickly got started on a yard sale at the farmhouse. I made a couple trips a day, each with a carload of our belongings to the new house. Gene and Amanda stayed behind to work the yard sale. Our move went smoothly, and we settled into our new home by mid-July of 1997.

The other members of our small family now included our pets: Thumper, the bunny; Cookie, our Persian; Trixie, our English Springer Spaniel; Charolette, the tarantula; Rainbow, the Beta fish; and Gary the snail. (Gary has since passed away.)

Our little family enjoyed the country living only a few miles south of a small community called Lindsey. Lindsey's general store was convenient in case we ran out of a certain something, or needed a fill up on gas. Lindsey Bar and Grill serves good food and offers live musical entertainment. The Lindsey park is a quiet place to hang out. And Hewitt's Meat Processing, owned and operated by Gene's cousin, is well known for their variety of tasty meats.

Life went along quite normally for the first seven

years our family lived on the wooded ten acres, until one night in mid-October 2004, when my young daughter and I were returning home from Marshfield with Halloween pumpkins. We were suddenly taken aback by the sight of something truly out of the ordinary. After that, things didn't feel normal anymore. We needed an explanation for the things we were seeing. But, it turned out to be a bigger mystery than we could have ever imagined. Did we see something we shouldn't have? Or was it meant for us to see? Will we ever know?

1

Do You See Those Lights?

"Do you see those lights?" I asked my eleven-year-old daughter who was sitting in the backseat. "Do you?"

Amanda propped her elbows up on the seat and peered forward. "What lights?" she asked.

"Up there!" I said, pointing ahead of us. "Are those lights in the sky?"

"Yeah, they're in the sky," she answered quite certain.

"Well, what the heck is that? It's just sitting there!"

I said, with my heart beginning to pound. This all seemed very unusual. I then began thinking, "Where have I seen this before?" The lights just sat there, motionless, stretched from tree-line to tree-line, above the center of the road!

We were the only car on Highway N, heading west toward Lindsey, when suddenly a car was coming towards us and I dimmed my lights. Then, I quickly returned my gaze back to the sky directly in front of us. And there, about a half mile or so ahead of us, hovered a formation of about ten white 'lights' the size of golf balls, in the shape of a sideways V!

"Do you see those lights, just sitting there?" I asked her again. It felt as though I was being hypnotized by the sight of them!

Suddenly, Amanda yelled, "They're turning off!"

I slowed way down so I could get a better look at this thing in the sky. And sure enough! The lights were shutting off! One at a time! Starting from the left, to the midpoint of the V, and then up the other row! I just couldn't believe what I was seeing!

"It was like, *bing, bing, bing*, and then they were gone!" Amanda exclaimed, staring out the window.

My thoughts were racing! What was that? It was there, and then it was gone! We both just looked at each other. "Where the heck did those lights go? They have to be close, they couldn't just drop out of sight!" I thought out loud. We slowed down even more to search the sky all around us as we continued on our way home. We couldn't see anything beyond the tall trees. "Wow, I think we just saw a UFO!" I gasped in disbelief.

"They're not going to come back for me are they?" Amanda asked, leaping into the frontseat with me.

"What! No! Are you kidding?" I chuckled, wondering why she said that. "This is a once in a lifetime event! We will probably never see anything like it again!"

Boy, was I wrong! This was only the beginning!

As we turned the corner onto Cary Rock Drive, there it was! Another light show off in the western horizon, on the tree-tops! "Twice in one night! Now, what the hell is that!" I exclaimed, as we stopped on the road in front of our house. "It's like a big ball of flaming light bouncing across the tree-tops!" I said. "The trees should be on fire!"

"It keeps bouncing back and forth!" Amanda replied. "Look, now it's just sitting there!"

I put the car in park and shut off the lights. "Let's watch this thing," I said. Suddenly, something small shot straight up from the light! As we watched, it stopped in midair! Then, it started moving straight to the right, and stopped when it reached the tree-line on the side of the road! All at once, it went straight left! We watched as it reached the southern sky. I'm not sure how long we sat there in awe, as this small light moved without flashing across the sky in front of us. The small light kept moving south, but the big bouncing ball of light was gone.

"What the heck is going on here tonight?" I asked, as I pulled the car into the driveway. After we unloaded the groceries, we walked down to the road and checked the skies, but, saw nothing! Feeling confused, we ran back to the house.

Okay, so that was the strangest thing we ever saw in the night sky! We couldn't wait to tell Gene, Amanda's dad, when he arrived home from work at eleven-thirty. As soon as he walked through the door, we both started yelling over each other about the lights we saw in the sky over Highway N, and toward the west, on the horizon!

He didn't say much. I guess it took him by surprise. But, Amanda and I were really pumped up about it! I even thought about it in my sleep. I knew what I had to do!

The next day, I visited the library and checked out some books on UFOs. I picked the ones with the most pictures. When I got home with the books, Amanda and I paged through them quickly, looking for pictures of what we saw. We both got goose-bumps when a picture of the V-formation of lights turned up on the page in front of us! That's where I saw this before! In UFO documentaries! My heart started pounding and I felt strangely weak.

"There it is, that's what we saw! Oh my God, we saw that! Exactly that! A UFO? By Lindsey?"

TIME AND DATE OF SIGHTING:

Between 8:00 p.m. and 10:00 p.m.

Mid-October, 2004

2

Some Strange 'Moon'!

WHILE AMANDA'S DAD is at work in the evenings, we usually go to Marshfield and run our errands. This particular night, we arrived back home around eight-thirty. As Amanda carried some groceries up the steps, I reached into the car for more bags. When I stood up and turned for the house, something caught my eye. I looked up, and through the leafless trees to the northeast, I clearly saw something. "What's that?" I asked myself. "Was that there a minute ago?"

Just then, Amanda came out the door to retrieve more bags from the car. "Hey, what's that light up there shining through the trees?" I asked her, with hopes of a quick, positive answer.

"I don't know," she replied, looking somewhat puzzled. "It looks like the full moon, don't it?"

"It looks more like a spotlight or something! Why is it so reddish-orange?"

We both stood there and peered over the car roof at this light. It looked so strange for some reason.

"Is that the moon?" I finally had to ask. I couldn't tell for sure.

"Why wouldn't it be?" Amanda turned and looked at me. "It looks like the full moon."

"I can't tell through the trees. It seems to shine different! Don't you think?" I asked, beginning to feel a bit uncomfortable.

"Well, can't you tell if that's the moon or not?" Amanda demanded.

"Not really. Hurry up! Let's get the flashlight and drive down to the corner!" I started the car as Amanda ran to get a flashlight. When she got back, she hopped in front with me.

As we approached the highway, about one hundred feet from our house, we saw what looked like the full moon hanging over the highway. It was at tree-top level and to our left. As we both stared at this 'moon,'

I noticed a strange red-orange haze, or dust, falling ever so slowly from the air all around it! I gazed out my driver's side window, trying my best to identify what it was we were looking at.

"Is that dust from Mount St. Helens erupting lately? Could the moonlight be shining through to make it look like this?" I asked in a whisper.

Just then, Amanda propped the flashlight on the dashboard and was about to turn it on when I yelled, "No! Don't turn that on!"

Looking startled, she froze. "Why not?" she quietly asked. "That is the moon, isn't it?"

"I think it is, but just in case it isn't, don't shine that light at it! Okay?" I said, as I gazed back at this 'moon.'

"Why? What could happen?" Amanda asked.

I wasn't sure what to say next. This 'moon' thing was so close, like you could reach out and touch it! In some UFO books I just read, strange things could happen to people after they flashed lights back at 'unidentified lights.'

But, this must be the moon! So, nothing will happen if we flash a light at the moon! How could I ever think that this glowing object, hovering above the road, wasn't the moon? Now, that would be crazy!

"It's following us!" Amanda yelled, as I drove forward to turn around in a neighbor's driveway.

"The moon always looks that way when you're moving," I reassured her.

But, by now, this 'moon' looked as though it had moved east, from above the road, to a small grove of trees across the road. It was almost like it moved with us! I'm not sure how, but suddenly, we noticed a gathering of coal black clouds just to the right of it. Some of the clouds stretched out over the top of this 'moon' like two fingers. Some of the clouds reached northeast towards Marshfield. At the time, the sky behind us was almost clear. This was very strange indeed!

As we drove back to Cary Rock Drive, we were still trying to decide why this looked so bizarre! We pulled the car up to the house slowly. We didn't take our eyes off of it as we climbed the steps to the house and closed the door behind us.

Amanda hung her jacket over the kitchen chair and just stood there, staring.

"Well, good thing we found out it was only the moon," I said slowly, with some uncertainly.

"Yeah, good thing," Amanda sighed.

Although we still had some doubts, we must have decided it was the moon. After all, what else could it be? It was only common sense! Right?

From my bedroom window I could see its circular glow through the trees. It was just enough to know it was still there. I checked often until I went to bed. I fell asleep thinking about how dumb this all seemed. My imagination must be running wild!

That weekend, Amanda and I decided to draw pictures of what we saw and show them to Gene. As we discussed this queer incident, it finally dawned on me! That's it! Why didn't I think of it right away? Just check the calendar! And, as I counted the days back, I discovered the full moon was ten days ago! The moon is no longer completely round. What we saw was completely round! And, it was a solid color of yellow-white. No craters, no shadows. And that funny red-orange dust surrounding it!

"Well, that solves it! It couldn't have been the moon!" I said for certain. "Right?" I then asked, looking at Gene and Amanda. "We should know what the moon looks like, and not be so undecided! We had plenty of time to look at it. It was right there close to us." Too close! It gave me an unsettled feeling, and not a day went by after that when I didn't second guess what it was we described to Gene that weekend.

Why couldn't I just forget about these lights we've been seeing? I know people will think I'm crazy if I ask them if they've ever seen UFOs around the Lindsey area. But, so what! We saw what we saw, and drew pictures of it. What if it wasn't the moon? Could it have been another UFO? We saw something 'unidentified'!

TIME AND DATE OF SIGHTING:

Approximately 8:30 p.m.

November 4, 2004

3

Read All About It

BY THIS TIME, I thought I read my share of UFO material, but apparently there was more I needed to learn. Because of our recent sighting of the 'moon,' I needed to learn as much as I could about the moon's phases. So, I headed back to the library. That night, I met Tom, a young man who must have also been interested in the paranormal because he drifted down the bookshelf in my direction. "Excuse me," he said, as he reached in front of me to pull a book from the shelf.

"UFOs!" I said. "Why, have you seen something?"

"No," he replied with a grin. "But, I've been interested in UFOs since I've seen E.T."

As I stood there flipping through the pages of another UFO book, I had the urge to say something to him about the lights my daughter and I had seen. But, what if he thinks I'm crazy? The 'moon' we saw just a few days ago could have been the moon. So I would feel really stupid bringing all this up. But, then again, what if he knows something that could help me? I thought for a moment and then just blurted out, "You saw E.T.?"

"Yeah, you know, *E.T.* the movie," he answered back, still grinning.

"Oh yeah, *E.T.* the movie," I laughed. How could I think he had meant anything else? I decided to tell him what my daughter and I had been seeing. He seemed quite interested and began to ask questions. So, I just continued to ramble on about our experiences. He actually believed me, and said he could look up some sightings from "UFO Wisconsin" on the internet. He would copy them and send them to me because I didn't have a computer. He also knew of a paranormal investigator he could e-mail.

"That would be great!" I told him. I gave him my address and thanked him for his help. As I left the library, I thought to myself, "I can't believe he didn't think I was crazy."

It wasn't long before Tom sent me the first proof of UFOs being sighted in neighboring towns. Could I believe what I was reading? The amount of sightings were unbelievable. So does that mean that what we saw was real?

I began to check out UFO documentaries from the library and watch them over and over. I also read every book the library had available on UFOs and related subjects. I compared what we saw and what other people saw. I came to the conclusion that something must be out there, because we were all seeing the same things!

So when Amanda and I started seeing very strange lights by Lindsey, we couldn't ignore it and pretend it didn't happen. We were not going to hide our thoughts so that people wouldn't think we were crazy. Besides, I doubt that other people who say they've seen strange lights in the sky believe that they're crazy!

Well, needless to say, Amanda's dad had not seen anything. I sometimes got the feeling he thought I was making too much of a deal out of this 'light' thing. But, it just so happened that it all changed one night in mid-June. I guess that's why they say seeing is believing!

4

They're Back!

As the three of us were driving back home from Marshfield on a Thursday night at about ten p.m., another light display took us completely by surprise. As we approached the top of the hill, I noticed some deer in the ditch to my right. I told Gene he should slow down in case they decided to cross the road. We were just south of Lindsey, about three miles from home. Then, all at once, this big, bright, white light appeared to our left. It looked as though it was situated just above a lone tree at the end of a neighbor's driveway. This thing was huge! It was like a basketball in the sky! Within seconds, a smaller white light, about the size of a baseball turned on to the right of the big light. Then, another baseball-sized white light turned on next to the first two. Now, Gene and I were seeing three white lights at tree-top level to our left! Will there be more?

I pushed myself back in my seat from total fright! Is this the V-formation that Amanda and I saw in October? If it is, it's going to be huge! I couldn't help but freak out! I started screaming. "It's back, they're back. Oh God, it's back!"

As I went hysterical, Amanda sat up from the back seat and asked what I was screaming about. At that very moment, the lights began to shut off, one at a time, in the opposite direction they came on. Then they were gone. Gene turned and looked at me and said, "I just got goose-bumps up and down my neck and back!"

"See, see, do you see what I mean by lights turning on and off?" I said, practically in tears.

No need to mention, we were all just a little shook! The three of us began to search the sky, wondering where it went. It was so low, yet no sound. What were those lights doing only a stone's throw south of Lindsey?

I don't recall how fast we were going when we noticed the first light suddenly flash on beside us, but it seemed as though we were standing still.

As we drove along slowly for home, we looked around the sky in all directions for any sign of a light, but we saw nothing. I just can't believe that any of our man-made aircrafts would flash three bright lights at a passing car, turn them off one at a time, and then completely disappear from sight. And why would an aircraft fly around without their lights on? Besides, the lights appeared to be so big that if it was an aircraft, it would have had to have been very close to the ground.

"Well, so what do you think now?" I asked Gene as we pulled up to the house. "Where the heck do you think something like that could have disappeared to?"

"Damned if I know! I would have stopped and looked around a little closer, but something told me to just get the hell out of there," Gene replied, with a tone of voice I only detect when he's flat out serious.

"And it was too bad I didn't see the whole thing!" Amanda added. "I just noticed the last light shutting off when Mom started screaming."

"Yeah, I was screaming because I thought it was that V-formation of lights we saw in October. If it was, it would have been huge, and right above the car. But, maybe I scared it away when I started freaking out!" I said to Amanda and Gene. Such a thought gave me a very weird feeling.

"Well, we can't be sure what it was because it all happened so fast." Gene added. "At least we got the hell out of there and made it home!"

All of us definitely agreed with that thought. Now finally, Amanda and I weren't the only ones in the family to see strange lights in the sky. Maybe now Gene had a feel for what was going on when Amanda and I were freaked out by what we saw in October and November.

So from now on, I'm sure that Gene will never second-guess what it is we described to him. I think he'll be wanting to watch the sky more closely from now on.

Maybe I will need to ask our neighbors if they have witnessed any of these strange lights themselves. That is, if they don't think I've gone crazy. We all know that seeing is believing. But, so far to us, the lights remain a mystery and 'unidentified.'

In the mean time, I continue to gather endless information about what people all over the world call UFOs. Therefore, the mystery of the Lindsey lights deepens!

TIME AND DATE OF SIGHTING:

10:00 p.m.

June 16, 2005

5

When You Least Expect It

FOUR OR FIVE months passed and we did not see lights in our night skies, except for the ones that belonged there. But, we all had the fever! I guess I had it the worst of the three of us. It's the 'fever' of needing to endlessly search the sky for lights, though after a while the searching seems fruitless. But, as the saying goes, things happen when you least expect them to happen!

Just as it did one night in October or November. (I must have forgotten to log this sighting on the calendar after it happened, so the exact date is unknown). Amanda and I were going into Marshfield to swim at the Senior High School pool. I was putting our backpacks in the car and turned around to head back to the house for my purse. When Amanda, who was standing on the deck, started yelling, "Lights, lights!"

I spun around to follow her pointing to the southeast sky. And there, right before us, was a row of four to six whitish lights, each the size of a softball! They seemed to come on towards us! Then, they went off, one at a time, in the opposite direction! All we could do was stand there and watch!

Then, a few seconds after that, not just one, but two rows of lights lit up, one at a time! We were then seeing two rows of yellow-white lights parallel to each other! The end of one row extended a little farther than the one beside it. All the while, they remained stationary.

As we both just stood there, clinging to each other, we watched as the lights disappeared and reappeared about three different times in the same place! We didn't know what to do! We couldn't move! We just needed to watch them.

"Call someone!" Amanda cried.

"Who?" I replied. "By the time I call, the lights will be gone!"

"Then, take a picture," Amanda suggested. But, just like the snap of your fingers, they disappeared and were gone!

"Let's go!" I said quickly. Amanda jumped in the car as I went back into the house for my purse. Deciding to take the camera along too, I grabbed it from the top of the microwave. But, before I could get to the door, the car horn was blowing.

"Lights, lights, they're back, they're back! Hurry up!" Amanda yelled.

I threw open the door, and got a glimpse of a row of lights, just as they were turning off! I ran to the car where Amanda was waiting in the backseat. "Did you see 'em?" Amanda cried. "They turned on right in front of me! They freaked me out! I was out here by myself!"

"Yeah, I know! Sorry it took so long. I decided to bring the camera. Let's stop on the road and see if we can get a picture of them!" I said, jumping in the car, half out of breath. I backed the car out of the driveway, and we waited on the road a few minutes, but nothing happened.

"Okay, now they don't turn on just because we have the camera!" I said anxiously. We kept looking around, and then noticed two sets of small white flashing lights, like planes, in the area the lights disappeared. "Those planes must have seen those lights, huh?" I asked.

"Yeah, you would think so, wouldn't you?" Amanda replied.

"Well, let's get going, and keep our eyes on the sky," I said, stepping on the gas, and pulling onto the highway.

The drive to town and back was uneventful. The sky was clear and quiet. As we got home, we remembered what we saw only hours ago from our deck. We started thinking that maybe those "planes" we saw after the lights had disappeared, really weren't planes!

Well, I guess we will never know. It seems to happen when you least expect it!

Or does it?

TIME AND DATE OF SIGHTING:

About 6:00 p.m.

October or November, 2005

6

'Sparkles' and Shine

As I set the table for dinner, I quickly gazed over at the clock. It was almost six p.m. That was about the time we saw lights last night! "Amanda, if you want to start eating, go ahead. I'm going to run outside and check the sky for lights!" I grabbed the flashlight, and dashed out the door.

It was cloudy, and no stars were visible. As I got to the end of the driveway, I thought I saw something! And there they were! 'Sparkles,' all over the southern sky!

They were on, or in, a low-hanging cloud. I couldn't believe what I was seeing! The whole southern sky sparkled with lights! They couldn't be stars, it was too cloudy. They couldn't be planes because they seemed to stay in one place. They just sparkled back and forth at one another.

I turned around and ran for the house! As I busted through the door, Amanda turned from her dinner and just looked at me. "Hey, there's all these little 'sparkles' flashing in this cloud! I don't know what they are!" I said, catching my breath. "They're like stars, flashing back and forth at each other, and there's a lot of them. Hurry up! Come with me and see if you can tell what they are!"

Amanda grabbed her jacket and ran with me out the door. We only went halfway down the driveway and stopped. The sight of these 'sparkles' was almost eerie! Sometimes, they seemed to form shapes as they sparkled back to one another. There must have been ten or fifteen of these very tiny lights. A few would sparkle from my left, and then a few would seem to answer back from my right. This just went on and on for quite some time.

"I'm going in," Amanda suddenly said, as she turned and ran off to the house, and left me standing alone.

"Hey, don't you wanna watch this?" I yelled after her.

"No, I seen 'em! I think they're weird and creepy!" she answered, running up the steps.

I turned back to the sky. The flashing continued. "What the heck could those lights be from?" I asked myself. This was like nothing I'd ever seen before! I couldn't make any sense of it! I felt chilled, so I ran back inside.

Amanda was just finishing her plate when I grabbed some chicken for myself, and tried to concentrate on eating. My mind was still trying to rationalize what I just saw in the sky, by our house!

"Are they still there?" Amanda asked, breaking the silence.

"Yeah, but they did seem to be moving more over the neighbors barn," I said with a sigh, as my mind flooded with deep thoughts.

"Do you want to take the car, and see if we can follow them?" Amanda suggested.

"Sure, let's do that," I answered quickly. "You grab the flashlight and I'll get the car started." I dashed out the door. "Check what time it is first!" I came back in yelling.

"What? Why?" Amanda asked, looking at me funny.

"So we know if there's any missing time!" I answered, with all seriousness.

"Okay? It's almost six-thirty," Amanda replied.

We hopped in the car and drove south down Lindsey

Road. There seemed to be only a couple 'sparkles' we could locate. We drove for a few miles, but decided to turn around when we couldn't find them in the dark sky. It's like they just drifted out of sight.

"Well, that's two nights in a row," I told Amanda, as we pulled into the driveway. "What are we possibly going to see next?" And when? I guess we just need to keep watching!

TIME AND DATE OF SIGHTING:

About 6:00 p.m. to 6:30 p.m.

Mid-October or November, 2005

7

Big Red!

BY THIS TIME, our sightings were really adding up! I had been keeping track of them on a calendar. One day, while reading the paper, I ran across an article about the paranormal, which included information about a book signing at a local bookstore in Marshfield. In the article, I found the name of someone who sounded really interested in strange lights. I looked her name up in the telephone book, and gave her a call. I introduced myself and explained my reason for calling. Diane became very helpful in finding me more information about UFOs on the computer. And she found plenty!

It had been a few months since we had seen lights, but we always checked the sky when we drove at night. This one night, I left for town alone. Amanda stayed home with her dad. He was home during the evenings at the time. I had no idea that what I would see that night, would turn out to be the most frightening sighting of them all!

I finished my business in town, and headed home the usual way. I was just a few miles out of town, heading south on Lincoln Avenue, when I thought I saw 'sparkles' in the southern sky!

I was the only car on the road, so I decided to pull off onto the shoulder. I turned my lights off and peered into the southwest sky ahead of me. And there! It looked to be real faint 'sparkles.' "It's true! I think I see something!" I said, under my breath.

I decided I better get moving in case a car would come, so I turned my lights on and pulled back onto the road. I hadn't gone more then a mile, when, all at once to my left, and at the top of my windshield, appeared a single bloodred light, about the size of a softball! It was there for only a flash, then it was gone!

"Wow, what the heck was that red light? Something

is out there! What should I do?" I said out loud. I kept driving slowly, while keeping my eyes glued on the sky around me. "I just wanna get home, I just wanna get home!" I thought out loud.

I began to feel as though that red light was meant for me! I started shaking, and my hands felt sweaty on the steering wheel. I needed to see what was out there, so I pulled over again, and shut off my lights. As I sat there in the dark, I felt as though something was there, but I wasn't seeing it!

"Okay, so it must be gone!" I decided. I pulled back onto the road again, and continued on my way. "About another fifteen minutes and I'll be home," I whispered to myself. I continued to drive along slowly, watching the ditch for deer, and at the same time watching the sky for lights. Every few miles I would stop where there weren't as many trees, so I could see out in the sky for a good distance. I reluctantly, but anxiously, checked all around. I saw nothing but stars. "What if it's right above me?" The thought gave me goose-bumps!

I must have stopped about five times in all, but never again saw the red light. The last time I checked was on Highway V, about two miles from home. As I looked real closely at the southwest sky, I thought I saw 'sparkles,' but I wasn't sure. Maybe by now, my eyes were playing tricks on me.

"Oh well, maybe it was nothing!" I said out loud.

I turned my lights on once again, and crept back onto the road. My mind shifted briefly to another thought . . . "Ah, a light! Oh wow, lights!" I suddenly screamed. "Lights!" Then, I saw another red light, to my left, at tree-top level! Across the road from our house! By the time I saw three red lights the size of softballs, I was beginning to get quite scared. Time seemed to stand still. I couldn't get the car to go fast enough!

I began blowing my car horn! The lights seemed to come on even faster! I was now seeing about six red lights in a row. A big row! Almost as long as a barn! The lights came on in the direction I was headed. They're right across from our house! Gene and Amanda! Do they know this thing is there, so close to them?

"Oh please," I thought, "make this thing go away!" I laid on my car horn without stopping as I passed a couple neighbors. "Please, someone see these lights!" I begged. As I finally approached the corner of Highway V and Cary Rock Drive, I couldn't bear to look at what was hovering above the trees alongside me! I rounded the corner, blowing the horn, never once looking over my shoulder!

Finally, the driveway was in view. As I whipped into the yard, I couldn't believe that no one was outside! Didn't they hear my horn blow? I slammed the car into park, and waited for someone to come running out. No one did, so I blew the horn again! I was too scared to get out of the car, and too scared to look for the lights.

Then finally, the door opened, and Gene stepped out. I threw the car door open, jumped out, and slammed it shut behind me. I ran up the steps, screaming and crying. "I saw lights, by Marshfield! I wasn't sure, but I think they followed me! Now they're out here!"

"Where?" Gene asked, looking around me.

"No! Don't go out there!" I yelled, as I pushed him back inside.

Amanda looked over his shoulder. "What's happening, did you have an accident?" she asked.

I shoved my way into the house trying to catch my breath. I nervously paced back and forth across the floor as I cried over and over, "I'm not kidding, I saw red lights! They were huge! What do they want right across from our house?"

I felt sick to my stomach, and I doubled over in pain. I was so upset I didn't know what to do! The groceries were still in the car, but I was too scared to let anybody go outside.

Gene opened the door and peeked out. "I don't see anything out there," he said.

"You don't know that," I replied.

"I'll go out and get the groceries," Gene offered.

"And I'll take a look around."

"Wait! I'll go with you," I said quickly. "It could still be there! Then what?"

"We'll all go," Amanda suggested.

"Okay, we'll all go, but let's put Trixie on the leash first. And grab a flashlight!" I added. We all walked slowly down the driveway. "It was right over there!" I cried, pointing to the east.

"Look, the 'sparkles' are there!" Amanda yelled.

We watched them as we unloaded the car. Then, they just drifted away. Until we find out what has red lights like that, they remain 'unidentified'!

TIME AND DATE OF SIGHTING:

From about 8:30 p.m. to 9:30 p.m.

February 28, 2006

8

One Starry Night

After what I saw the night before, I didn't feel comfortable driving alone after dark. So, tonight we all drove together to Marshfield. On our way, we watched for lights, but we saw nothing out of the ordinary. The following night, we took another ride to town, mainly to look for lights, I guess. We drove a different road home that had a better view of the southern sky.

We saw a few lights that, at first, looked suspicious,

but were only planes. We did notice a great deal of bright stars located in the southeast.

"Look at all those stars!" Gene commented, as we reached the end of the road and turned west.

"Yeah, wow, cool stars!" Amanda and I agreed, as we took in their beauty.

We were just a few miles from home now, and were a little disappointed that we didn't see anything worthwhile. As we pulled into our driveway, and parked the car, the southern sky was lit up with all those stars! While I lifted a bag of groceries from the trunk, I gazed back over my shoulder. Those were the brightest stars ever! For some reason, they made me feel good that they were there! "Wow, would you look at those stars!" I said to Amanda, as she came back for more bags.

"Yeah, they're really bright! They're lighting up the whole sky around them! And see, there is a bigger one almost in the middle of them all!" Amanda commented.

We both just stood there a few minutes, and gazed up at the stars. I closed the trunk of the car, and we walked to the house.

Before the groceries were unpacked, I decided to go back outside and get another look at those brilliant stars. When I reached the end of the driveway, and still couldn't see them, I thought for a moment. "Where the hell are all those stars? What the heck! Where did they go?" I ran back to the house, and busted through the door yelling, "Hey, those stars are gone! I don't see them at all! Nowhere!"

Gene and Amanda both looked at me. "Maybe the clouds covered them up," Gene suggested.

"No!" I insisted. "There aren't that many clouds! The other stars are out!" For a moment they both just stood there, and looked at me like I was crazy. "Hey, what if they weren't stars! Don't you think they looked different for some reason?" I asked, starting to feel really confused by what we just saw. But, what else could they be? They looked like stars!

"Well, they did look different," Amanda finally agreed.

"There sure was enough of them." Gene added.

"Yeah, they lit up the whole sky around them! Let's go see!" Amanda said. We both dashed out the door. We took a quick look around the sky and ran back to tell Gene. "Yes, they're gone alright!"

As we stood around talking about these silly stars, I realized another strange thing about them. They all seemed to be of the same magnitude and size! Maybe what we saw really wasn't a thick blanket of stars! What now? UFOs that looked like stars? Or, maybe a strange constellation of traveling stars?

I know that I would recognize them, if I saw them

again. Besides, if they were truly stars, they would be in the same place at the same time tomorrow night. We all agreed on that!

The following day, I couldn't wait for eight-thirty to come around, so I could look for those brilliant stars, and put my mind at ease. But, throughout the day, I couldn't help thinking, "Those 'stars' are not going to be there!"

Finally, eight-thirty arrived, the same time as the night before. I ran out the door, and down the driveway. I scanned the sky as far as I could see. I had to be absolutely sure! The 'stars' were not there! Just as I thought! Do we really need to watch the sky every night! Because there seems to be no end to the strange and bizarre lights in our skies, by Lindsey, Wisconsin.

TIME AND DATE OF SIGHTING:

Between 8:30 and 9:00 p.m.

March 2, 2006

9

Jet Action

IN SOME OF the UFO material I had read recently, some people described hearing the sound of loud military jets at the same time as a UFO sighting. By this time, we had witnessed several different types of lights. Some were at tree-top level, and some were out more, above the horizon. The colors of some lights were white or yellow, while some were bloodred. Still, others resembled celestial bodies. But, all were seen between Marshfield, and our home. Most of them right from our yard!

One evening, about three weeks after our last sighting, we were all watching a movie, when I heard military jets overhead. I jumped up, and ran to the south window. Looking out into the darkness, I thought I saw a flash of light, above the tall trees in front of our deck. I quickly stepped outside to get a better look. I could hear loud military jets, but couldn't see them. When suddenly, there was a flash of light by the trees, where I saw the flash just seconds before. And it was big!

"Hey, you guys, get out here!" I yelled through the open window. "There's a light out here! Hurry up before it's gone! I don't think it's a plane!"

Gene and Amanda jumped up, and came running, just in time to see a white light, about the size of a baseball, turn on above the trees. It seemed to pulsate, almost like a fluorescent light, before it came completely on. Then, all at once, a streak of light shot out of the left side of the big light, but it did not go very far. Then, within moments, the big light just went off, and reappeared in the spot where the small ball of light stopped! This pattern happened three or more times, as all three of us stood and watched in awe!

Meanwhile, we could hear the roar of a jet, and within seconds, it flew over the house and into the exact location where we had noticed this strange light beginning to flash! After the jet had passed through,

THE MYSTERIOUS LINDSEY LIGHTS

going south, we checked back to the east over our driveway, but the strange white light had vanished!

"What do you think that was all about?" I turned and asked Gene. "Could it really have been another UFO?"

"I don't know. But, I've never seen anything like that before! I bet those jets saw them on their radar, and they were chasing them!" Gene replied.

"Wow, yeah! Dad could be right, maybe the jets were chasing them. But, whatever it was, it was really weird!" Amanda added.

All three of us stood on the deck for a while, looking up into the sky. My mind kept rehearsing what I just saw. I wanted to think it was something logical, but then why would it seem so strange?

After we went back inside, and began to wind down from the excitement, I decided I had to do something to put my mind at ease. Tomorrow, I will find the number to Fort McCoy, which is about fifty miles south of us, down around the Tomah area. Maybe they could tell me if they were doing military maneuvers, and what I could expect to see. I felt that what we witnessed over our house was some really wild jet action. I needed an explanation. And, before I went to bed, I logged another sighting on my calendar.

The next day, I found the number for Fort McCoy. I wasn't sure what I would say, but I needed to go through with it. So, with my heart pounding, I picked up the phone, and began dialing. When the phone started ringing, I felt a rush of warmth all through my body. I told myself, "Stay calm."

After three rings, a female voice answered, and said, "Fort McCoy."

I responded with, "Yes. Um, I'm calling to ask about some lights we've been seeing by our house. I was wondering if you were doing maneuvers over this way?"

"Hold on," she replied. "I'll have to switch you over to infrared."

"Okay, thank you," I said.

The phone began to ring again, and this time, a man picked up. "Infrared," he answered.

"Yes," I began. "I live about seventeen miles southwest of Marshfield, in the Lindsey area, and for some time now, our family has been seeing, like, these strange lights in the sky, around our house. I was wondering if it was you guys doing maneuvers?"

"Okay. And what exactly are you seeing?" he asked politely.

"It's not like I want to know what it is you're up to, I was just curious what it would look like if you guys were doing military maneuvers, because we have seen different kinds of lights lately. For instance, there was this V-formation of about ten white lights that was hovering above the road. Then the lights just suddenly shut off one at a time, and then it was gone," I explained.

"Huh," the man commented.

"And we've seen other lights too. Some were yellow-white, and then, another time, I saw a row of about six red lights. These lights would turn on one at a time, and shut off in the opposite direction," I continued.

"Huh," The man commented again. "Well, it is open airspace out where you live. We do maneuvers over there. Could you hold for a moment please? I have another call coming in."

"Sure," I answered.

Within a few seconds he was back on the phone. "Okay," he said. "What were you saying?"

As I continued to describe what we saw, he just kept saying, "Huh." Then he asked, "Did you get any pictures?"

"No," I answered. "It always happens so fast that I don't have a camera ready, but I wrote it down in my calendar."

"I see," he replied. "I guess all I can tell you is that our jets have red and white flashing lights, and are easily recognizable."

"Okay?" I said, with question in my voice. I thanked him for his help, and said, "Goodbye." I couldn't tell for sure what he thought about all I had told him. The only clue he left for me was that their jets are easily recognizable.

So, if it's true that the military is flying in the open airspace by our house, that explains the jets we've been seeing. But, it doesn't explain the assortment of lights that have also been present. I got the feeling that he wasn't telling me something. And, of course, I didn't expect him to tell me their military secrets. Why would he want to know if I had pictures? And when he put me on hold, could he actually have been recording our conversation? He seemed pretty interested in my descriptions of the unusual lights.

Well, who knows what could happen from this point. I'm glad I made the phone call, because I wouldn't have known otherwise. And, maybe Gene was right. The military could have been chasing something.

TIME AND DATE OF SIGHTING:

About 9:15 p.m.

March 29, 2006

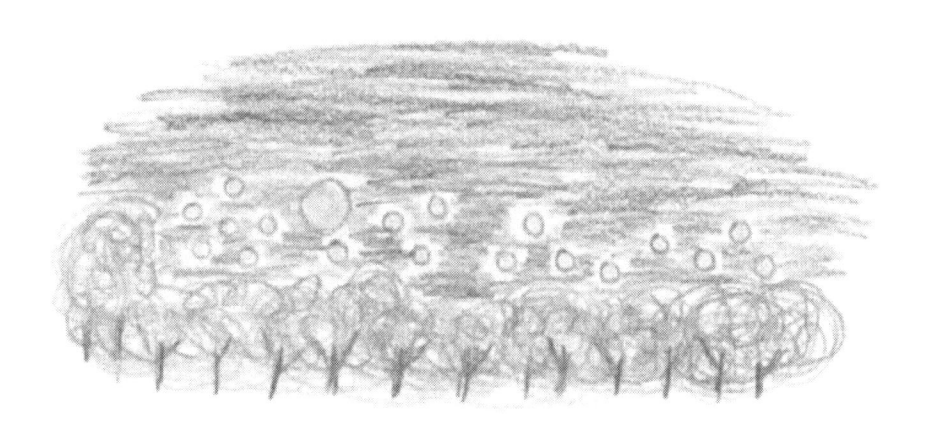

10

We're Surrounded!

WITH ALL WE had witnessed, I checked the sky almost every night after dark. But, like I said before, things happen when you least expect them to happen. Almost five weeks went by, before it happened again.

"Jets!" I yelled. I immediately jumped off the couch, and ran out the door. And there, in the southern sky, were two military jets, with red and white flashing lights. They weren't very high up, and were flying rather slow. I watched as they flew out of sight. "Now, those I could identify," I said to myself.

I stepped back into the house. Amanda looked up at me from the TV, and asked, "What's out there?"

"I just saw two military jets flying pretty slow, and heading southwest," I told her.

"That's all you saw then?" Amanda asked.

"Yeah, just jets," I replied.

A short time later, I decided to check the sky, just in case! I rushed down to the end of the driveway. It was cool out, and I didn't grab my jacket. As I scanned the sky all around, things seemed quiet. Then, all of a sudden, something faint caught my eye, in the southeast horizon, just above the tree-line. They appeared to be tiny, blue-white lights. As I watched, there seemed to be more of them, coming out from behind the taller tree-line! They didn't go far, but moved back and forth, flashing like 'sparkles.' "Are those jets?" I asked myself. "They don't seem to be moving across the sky!"

Feeling chilled, I ran back to the house to get a jacket. Gene had just gone to bed, but Amanda was still up. "I see some lights out there, I don't think they're jets! There's at least a half-dozen or more! I'm going back out to see if I can tell what they are!" I told Amanda, as I scrambled for a jacket and ran back outside.

As I got to the end of the driveway, it didn't take me long to decide they couldn't be jets! And, now there was more of them! Across the southeast horizon. Tiny, blue-white lights, each about the size of a star, flashing back and forth. I really couldn't make any sense of what I was watching! There was no sound. Nothing!

I didn't want to run back to the house to tell Amanda because I could miss something, so I yelled from the middle of the road, hoping to be heard through the half-open window. "Amanda! Hey! There's lights out here!" When no one came out, I yelled again, like a mad woman. "Hey, there's lights out here, hurry up! Amanda! Gene! Lights!"

Finally, they both came running down the driveway, and met me on the road, where I pointed to the southeast. "There, see that? They look sort of like the 'sparkles,' but much brighter, and they seem to sparkle faster!" I said.

As we stood on the road, watching these tiny lights randomly blink off and on, we began to notice some of them glowing brighter. Suddenly, one of them just exploded into a huge ball of light, and then simply shut off! Right before our eyes!

"Wow, what was that?" Amanda asked, with an unbelieving look on her face.

"I don't know!" I answered. "It just blew up! And there was no sound!"

The flashing would seem to slow down to where you could hardly see them. Then, they would suddenly appear to get brighter. All at once, one of them would seem to simply blow up into a big, bright light! "There it goes again!" I yelled. "What in the world is that? Those can't be flares! Flares don't turn on and off like that!"

"Should I go in and call Rodney, to let him know that we're seeing lights down here?" Gene asked.

"Yeah, go do that! But, hurry up! It's spooky out here!" I answered.

Rodney lives only a few miles east of us, so it's possible that he will be able to see the lights, and tell what they are. It wasn't long before Gene came running back down the road. "There was no answer," he said.

So, Amanda and I ran up to the house to phone my friend Diane. When she answered, I quickly explained what was happening. She said she would drive around, and see for herself. I hung up the phone, and we dashed back outside.

When we got down to the road, Gene said he saw the 'sparkles' blow up into huge, white balls of light, about three times, while we were in the house. "Wow!" we both exclaimed. "We missed it!"

We all just stood there, keeping our eyes on the 'sparkles,' wondering what the heck was really happening. We must have been on the road for at least ten minutes, watching the southeast sky, when Amanda yelled and pointed to the southwest. "Look, there's some over there too!"

Gene and I turned to the southwest. "Oh my God, they're all around us!" I cried. "What do you think they're doing? What are they?"

"Should I go get the binoculars?" Amanda asked.

"Yes, yes! Why don't we have the binoculars?" I exclaimed. "Hurry!"

Within seconds, Amanda came back with the binoculars and focused in on the southeast sky, where most of the flashing was taking place. Then, all of a sudden, another light blew up! "Oh, wow! That was just awesome! It was huge in the binoculars!" Amanda exclaimed. "Just huge!"

"Let me see!" I yelled impatiently. She handed me the binoculars, and I focused in on the same location, just in time to catch a light blow up big and white! Within a few seconds, another one blew up! I saw them both through the binoculars. "I just can't make it out! What the hell is all this! I can't believe what I'm seeing!"

I handed the binoculars to Gene. He stood there watching, as I paced the road. "There sure is enough of them," Gene finally said.

"I know! It's a whole fleet! What are they doing? This is getting scary!" I cried, almost in tears. "They're all around us!"

"We're surrounded!" Amanda yelled.

Then, almost instantly, they were gone! They just

stopped flashing! We stood there for a few minutes, waiting for them to reappear. But, nothing. "They must have left!" we all finally agreed. We walked back to the house, checking over our shoulders. When we got through the door, I looked over at the clock, it said nine-forty. We must have been out there twenty minutes or more!

Just then the phone rang. It was Diane. "Did you see it?" I asked, anxiously.

"No, I didn't see a thing! Maybe the hills were in the way. They must have been closer to you," she replied, with regret in her voice.

"Oh man, you missed it! We were out on the road for about twenty minutes! They were across the whole horizon! From the southeast to the southwest!" I explained. I was disappointed that she had missed the whole thing.

"Well, maybe next time," I thought. And there probably will be a next time! But, at this point, the lights were still 'unidentified'!

TIME AND DATE OF SIGHTING:

From 9:20 p.m. to 9:40 p.m.

May 4, 2006

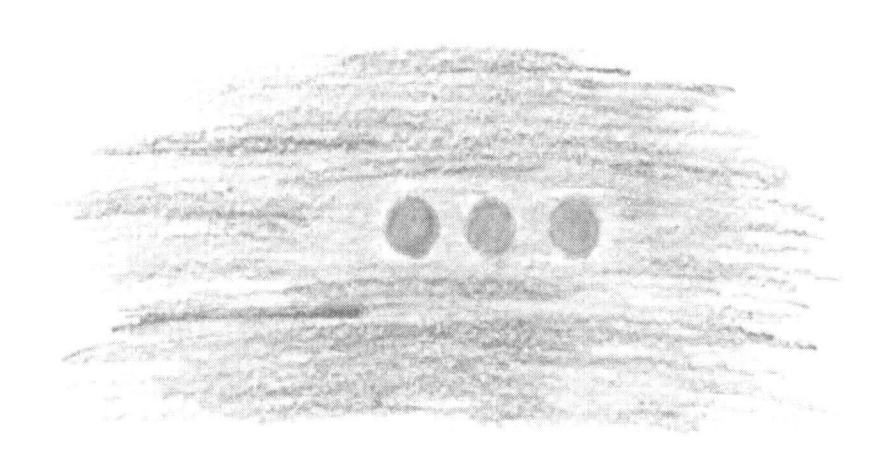

11

That Was No Plane!

"Can we look for UFOs tonight?" Amanda asked, with her hair still wet from her shower.

"Sure, I guess, for a while," I said, putting my book to the side.

"Good!" Amanda grinned. "I'll be ready in a couple minutes."

I got up, somewhat sleepy-eyed, and started gathering our stuff for our almost nightly ritual. Just as the sky would darken, and the stars would slowly appear, the UFO fever would strike! Amanda and I would sit in the car, across from our driveway, armed with a flashlight, binoculars, and our dog, Trixie, in the backseat. And there we sat, watching, and waiting.

"I see a light flashing," Amanda said, pointing to the southeast.

"Where?" I asked, trying to locate anything strange in the vast sky.

"Oh, I think it's just a plane. The lights are flashing red and white," Amanda answered.

"Yeah, yeah, that's just a plane then." I replied back. "There's another one over there," I pointed to the east.

"Yeah, they're planes alright. You can tell," Amanda said, looking through the binoculars.

As many nights as we spent on the road, looking for strange lights, you would think we were crazy. But, there is so much power in the 'fever,' that we just had to keep looking. Almost half an hour had gone by, and we saw nothing 'unidentified.' "Maybe we should call it a night," I suggested, feeling tired from a long day.

Just then, Amanda shouted, "I saw something flash back there!"

"Where?" I asked, yawning and looking through the back window.

"Over there! See, there it goes again!" Amanda yelled.

"Okay, now I think I see something flash!" I said. "But, let's make sure."

"There's more of them!" Amanda yelled, not being able to hold back her excitement. "There's more of them!"

"Wait now! Let's make sure they're not planes," I suggested, trying to stay calm.

"Mom!" Amanda yelled back. "Those are not planes! What we saw before were planes!"

As we watched closely, they seemed to be moving our way. But, then they traveled back toward the horizon.

"Now they look like they're coming back this way," I said. "But kind of slow."

"They are!" Amanda answered. "There's gotta be about ten of them or more. Those can't be planes!"

"Wow, do you think it's those 'sparkles'?" I asked quietly. They slowly made their way from the southwest to the south east, then, slowly moved back to the southwest. Tiny, single, blue-white, blinking lights, floating across the sky.

"You should go call Diane," Amanda suggested. "Hurry, before they disappear!"

I thought for a moment, watching the 'sparkles.' I didn't want to miss anything. "Yeah, okay!" I decided, as I started the car, and quickly drove up to the house. "Stay here and watch them. I'll be right back!" I yelled, as I rushed up the steps.

"Okay. But, hurry up! It's scary out here alone!" Amanda yelled behind me.

"I hope she's home," I said to myself, as I thumbed through my Rolodex for her number. I couldn't find it fast enough! Finally, I dialed the number with a shaky hand. The phone seemed to ring extra slow. "Come on! Pick up," I hummed impatiently.

"Hello?" a voice finally answered.

"Diane, it's Betty! You have to check the sky! There's a lot of 'sparkles' out in the south, moving back and forth! Amanda and I are watching them from the car, on the road! I gotta go! Amanda is waiting in the car, and I don't wanna miss anything! Talk to you later! Bye!"

I barely gave her enough time to say, "Okay." I hung up the phone, and ran back to the car.

"They're still there!" Amanda yelled, as I started the car, backed onto the road, and parked.

"What could those things be?" I asked. "Why can't we tell what they are?"

"They're not planes." Amanda said, with a sigh.

There was no sound whatsoever. And, they just kept moving from west to east, and back again. How odd.

"I wonder what would happen if we flashed the car lights on and off?" Turning the switch on, I gave it a try.

"What are you doing?" Amanda turned and asked,

somewhat startled.

"I just wanted to see what would happen," I answered slowly, keeping my eyes on the 'sparkles.'

"Well, something just did! It looks like they're blinking faster, and coming this way," Amanda whined.

"Maybe," I answered, flashing the car lights again. "I can't tell for sure."

"Are you crazy!" Amanda yelled back.

Just then, a row of three orange lights, the size of grapefruits, lit up, one at a time, in the southern sky alongside us! Amanda grabbed a pillow that was in the car, and propped it up in the window, peaking around it, at the row of three horizontal lights, hovering in the sky!

"Holy shit!" Amanda screamed. "Now I'm freaked out!"

As we watched, the lights went off, one at a time, and in the opposite direction they had come on! "Alright, now this is getting a bit more interesting!" I mumbled. "That was no plane!" I flashed my lights on and off one more time. I just couldn't resist the urge to see what would happen next!

"Don't do that!" Amanda screamed at me. "Are you crazy? Didn't you see those lights turn on, right beside us?"

Just then, a car came from the south, down Lindsey road. As the 'sparkles' continued above the moving car, suddenly, one flashed into a big, orange, ball of light, and then quickly shut off!

"What the heck?" I shouted.

"That light just reacted to that car coming down the road!" Amanda said, with true excitement in her voice.

"It sure seems that way! I just wonder if the people in the car saw it," I asked. "It was right in front of them!"

"I doubt it!" Amanda answered. "Who looks up in the sky? No one, but us!"

"I don't know. It was right above them! How could they not see it?" I wondered.

Within a short time, a car rounded the corner, and headed our way. "Is that Diane?" Amanda asked, as the car pulled over, and stopped just before our driveway. "It looks like her car!"

"Yeah, yeah, that's Diane!" I yelled, as we both jumped out of the car, and ran over to meet her. "Did you see them? Did you see those three orange lights turn on?" we each screamed over the other.

"I seen it! I seen it!" she yelled back. "I seen those three orange lights!"

"Alright! Alright!" we all yelled, jumping up and down in the middle of the road.

"Yes, yes, you seen 'em, you seen 'em! Now you know what I mean when we say we seen 'sparkles' blow up into huge balls of light!" Boy, was I glad

Diane didn't miss this one! Now, someone besides us three, have seen the strange lights in the sky, across from our house!

"Man, those orange lights that came on in a row, really had me freaked out! I almost drove off the road!" Diane exclaimed, with just a slight chuckle in her voice.

"Did you see the other one turn on just after that? We think it turned on because of a car coming down the road!" I told her.

"Wow! No! I was so freaked out by what I just saw, I didn't notice anything else! But, I did see those tiny 'sparkles'!" Diane said, walking up and down on the road. "What the hell is going on! That's not the military!" she said, as she looked the sky over.

"We don't know what it is, but we're glad you seen 'em too!" I replied happily.

All three of us watched the 'sparkles' drift back and forth for a while longer. But, like always, they just slowly faded away. We were still standing on the dark road, and staring into the sky, when Gene arrived home from work.

"What's up?" he asked, as he parked in the mouth of the driveway.

"We saw lights! Orange lights, in a row!" We yelled. We all started talking at once about the 'sparkles,' and big orange lights. "You missed it, they're gone! We've been out here for over an hour!" we said, as Gene looked the sky over.

The four of us then stood on the road, talking for a while about what it could have been that we just witnessed. Finally, Diane thought she'd better head home. So, we waved goodbye, and pulled our car up to the house. Of course, I logged another sighting on our calendar, and went straight to bed.

TIME AND DATE OF SIGHTING:

9:30 p.m. – 11:00 p.m.

July 18, 2006

1 2

Mysterious Lights

After all the excitement the night before, Amanda and I couldn't wait to watch the sky tonight! We gathered our stuff, and backed the car onto the road. There we sat, keeping watch in all directions. After about a half hour, I began to think we weren't going to see anything special.

"Well, should we go in? All I see are planes," I told Amanda, with a sigh.

"Can we look for just a little longer?" Amanda

asked. "We haven't been out here that long."

"Okay, maybe another fifteen minutes or so," I answered slowly.

As I settled back into the seat, and gazed into the night, my mind began to wander. What is really out there? Where is the life that may exist beyond our planet? I can't imagine we're all alone!

"The windows are getting foggy," Amanda said, waking me out of my daydream.

"Here, I'll roll my window down some," I said with a yawn. "But, I don't think we're going to see anything tonight. We've been out here almost an hour. I'm getting tired." It seems like we all need more sleep these days.

"Yeah, you're right. I guess we're not going to see anything tonight," Amanda replied, with disappointment.

"There's always tomorrow night," I answered, as I drove the car into the yard.

So, the following night, just as planned, when the sky was dark, we gathered the flashlight, binoculars, and leash for Trixie. Amanda stepped out the door before I had my jacket on. "There's 'sparkles' out there!" she ran back in yelling.

"What?" I asked.

"There's 'sparkles'! You can see them right from the deck! We gotta get going!" she yelled with excitement. "They're right out here!"

"Let's go!" we both said, grabbing our stuff, and leaping into the car. I swiftly backed the car onto the road. And there they were! About ten points of light, sparkling like snow in the sun. They drifted back and forth through the southern sky beside us ever so slowly. As we watched in awe, I couldn't resist flashing the car lights on and off a few times, just to see what would happen. It could have been our imagination, but we could have sworn that the lights moved closer, and started sparkling even faster.

"Alright, maybe I shouldn't do that anymore," I whispered, as Amanda turned and looked at me.

"Why?" she asked. "What do you think could happen?"

"Well, I don't know what they are," I said slowly.

"Could we get abducted?" Amanda asked.

"No! They're just lights! We haven't seen aliens!" I replied, not knowing what to really say. "Let's just watch them, and see what happens next."

"Well, they still creep me out," Amanda whined, looking through the binoculars. We both sat there silently for a few minutes, as the lights danced around the sky alongside us. "Mom, it looks like there's a ball of light on the neighbor's barn!" Amanda said suddenly.

"What? On the barn? Let me see!" I answered. Amanda handed me the binoculars. "I can't see it!" I said. Then, suddenly there it was! It wasn't very big, but I did see something that looked like it was on, or over, the barn. "What the heck! There is something there, but I can't make it out very good."

As we discussed this light on the barn, we noticed the 'sparkles' were drifting to the south, away from us.

"Let's drive down Lindsey Road, and maybe we can see the light that's over the barn better," Amanda suggested. "And maybe we can follow the 'sparkles.'"

"Yeah, we might as well," I agreed. We kept our eyes on the sky as we drove south on Lindsey Road. The light on the neighbors barn was nowhere to be found. We also noticed that the 'sparkles' had just dropped out of sight. We crept along for another mile, but lost sight of anything 'unidentified.' So, we turned around and headed home. "Well, we did see something over the barn, it wasn't our imagination!" If we could just once find out what these mysterious lights are, maybe then we could put an end to the mystery of the 'unidentified lights'!

TIME AND DATE OF SIGHTING:

Between 10:00 p.m. and 11:00 p.m.

July 20, 2006

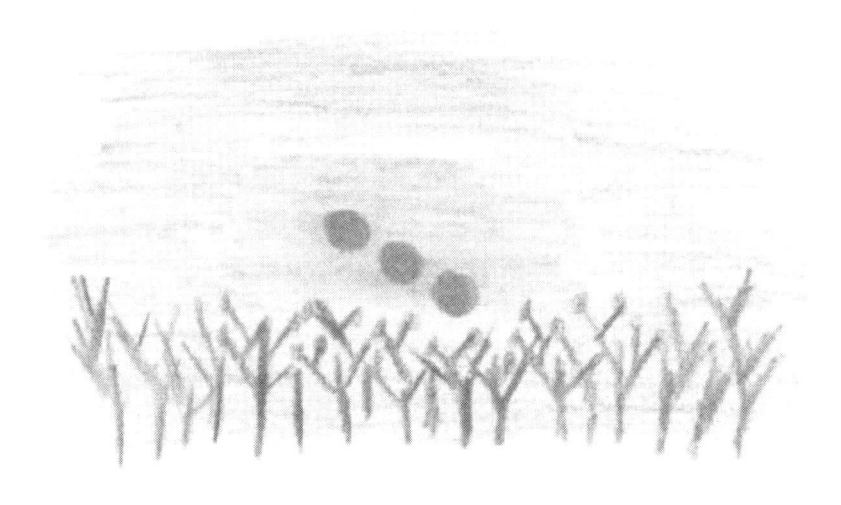

13

They're Still Out There!

By the fall of 2006, we had decided that our small house no longer suited our family's needs. Selling the ten acres was not an option. Building a new house was. The decision was quickly made.

We built our new house just six feet behind the old one. It was a two story, with balconies on the east and west sides, so this made a good lookout for 'unidentified lights.' For months, the sky had been unusually quiet, with no sightings of strange lights. Was the action over?

It was now mid-April 2007. I was leaving Marshfield alone, when the thought just happened to cross my mind that maybe I would see 'unidentified lights.' I was about five miles south of the Marshfield airport, on Lincoln Avenue, when *bing, bing!* On came two orange lights, one at a time, just above the trees to my left! There could have been more, but my view was blocked by the tall pines. The lights were the size of baseballs, and came on at an angle, and then quickly turned off. I was amazed! "I just seen lights!" I said to myself. I turned my radio down and reached over for my flashlight. I told myself to stay calm, not knowing what could happen next! I was the only car on the road again, as I approached Highway N. I knew what I had just seen was real! What should I do? What could I do? Nothing. Just keep driving, and watch the sky!

When I checked my rearview mirror, I noticed car lights coming a distance behind me. I was now heading west on Highway N, and although I like the road to myself, I was really glad to see another car! Keeping a steady pace, I watched the sky, and checked my rearview mirror often. The car was still behind me, and closing in. "Thank God I'm not the only one out here," I mumbled to myself. Then, I saw lights come up over the hill from an approaching car. "Oh good! Another car!" I exclaimed. I was feeling really anxious to get home!

Finally, I reached Lindsey corner, and headed south on Highway V. Three more miles, and I'll be home! I couldn't wait to tell Gene and Amanda that I saw two big, orange lights, just a few miles out of town.

When I parked the car, I sat for a moment, thinking about what I saw. As I got out of the car and opened the trunk, I took a quick look around the sky, and headed up to the house. I just wish I could be sure of what I saw.

"Oh my gosh Mom, you missed all the action! Dad and I have been seeing orange lights light up in the sky all night!" Amanda yelled down, from the upstairs window.

"What, are you kidding? I seen 'em too! A few miles from town! There were two, or maybe three! They came on and shut off quickly, so I didn't get a second look. You guys seen 'em out here too? Where?" I yelled back up to her.

"Over there, above the trees!" Amanda said, pointing to the southeast. "Too bad you weren't here!"

"Wow, I can't believe it!" I said. "There's no doubt in my mind that they're still out there!" I rushed into the house and met Amanda coming down the staircase. "So what did you all see?" I asked.

"Well, I was upstairs looking out the south window, when I got the feeling that I might see lights tonight. So, I got the binoculars out, and was looking at the stars when I saw flashing over the tree-tops. It looked like the 'sparkles,' so I called for Dad. By the time he got upstairs, the whole sky was full of them!"

"Wow, I didn't see 'sparkles'!" I replied.

"Then, the orange lights began to come on! Sometimes as many as six in a row! I don't even remember how they turned off. It all happened so fast!" Amanda rambled on.

"Your kidding!" I replied.

"Then, I told Dad to call Diane. But, there was no answer. So then, I said to call Gary. And, while Dad was on the phone, the 'sparkles' traveled around the sky, and the orange lights kept turning on and off at different angles. They were big, as big as softballs!" Amanda exclaimed.

"Boy, I wish I could have been home to see all of that!" I sighed.

"Then, a big explosion of orange light lit up a good part of the eastern sky above the tree-tops! And, in that glow, a row of orange lights turned on, going up at an angle from the tree-tops," Amanda continued. "Now that, was really freaky!"

So, after Amanda told me all that, we checked the sky all night long, but the mysterious lights had disappeared.

The following two nights were just as exciting. Amanda and I were on the east balcony, when suddenly, we saw 'sparkles'! And as we watched, the orange lights began turning on. Sometimes three or more in a row, evenly spaced! This continued for almost an hour. This was just amazing! I logged the sightings on my calendar. Three nights in a row! All about the same time of night. What was going on out there?

The next couple nights were quiet. But, the following week they were back! First, we heard jets. Then, as we watched from the east balcony, we began to see 'sparkles.' Does this mean the orange lights will be next? We remembered hearing jets at the time of the sighting last week. Could there be a connection between the jets and lights?

The following night, there was nothing. We assumed the excitement was over. But, we were wrong! That Wednesday night, just after dark, Amanda and I were again watching from the east balcony, when we began to see faint 'sparkles' in the southeast. All at once, we heard the loud roar of a jet passing overhead, going east. Then, another jet, with red and white flashing lights, flew over the house going south. All at once, the noise of the jet just ceased! We thought that was weird. Feeling scared, we rushed inside the house and took our post in front of the south window. There, we got a good view of the baseball-sized orange lights appearing and disappearing at different places in the sky. At one point, the orange lights appeared directly in front of us, forming the shape of a 7. They then turned off, one at a time, in the opposite direction. We looked at one another, and all we could say was, "Oh my God!" We noticed that the smell of sulfur had drifted in the open window.

In general, we don't understand what is really going on in our skies. But, some of us know that there are things still out there! And, if our government is involved, I guess they can fool most of the people most of the time, but some of us can't be fooled at all!

TIME AND DATE OF SIGHTING:

Approximately 8:30 p.m. to 9:30 p.m.

April 17–19, 23 and 25, 2007

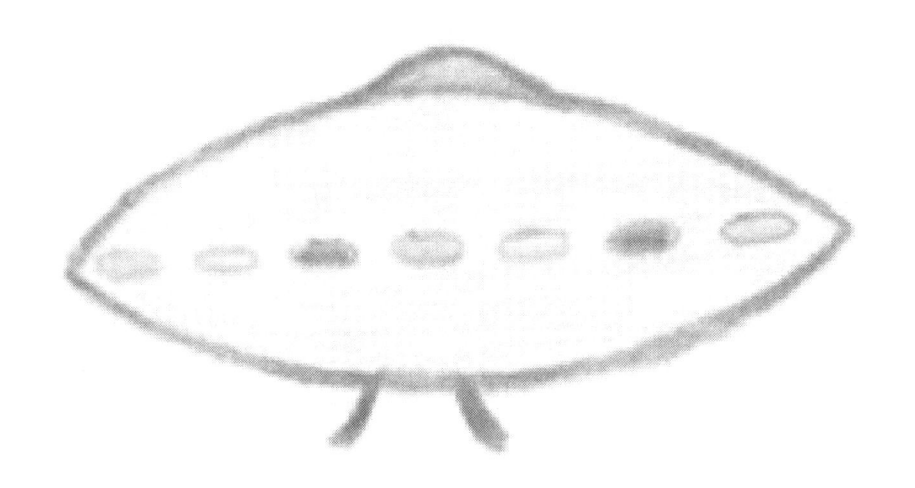

14

The Believers

OVER THE COURSE of the last few years, we discovered that some of our friends and neighbors had their own encounters with UFOs as well. Yes, they also saw strange lights, near the Lindsey area!

A young man and his dad, Gary, who both live between Marshfield and Lindsey, had seen an assortment of lights, that appeared to be hovering over their pond. When I spoke to this young man, he said he had witnessed lights hovering, and/or turning on and turning off, in somewhat the same manner as we had seen them. He was convinced that it was not the military, and considered them 'unidentified.' Because of the number of the sightings Jason had seen around the area, he now carries his camcorder with him, in the hope that he will be able to catch them on video.

Another witness lives just north of Lindsey. He happens to work with Gene, so I decided to call him and ask what it was he was seeing. He described to me what sounded like the 'sparkles' we had seen. He mentioned that he and his wife had witnessed what they thought was a daytime craft, moving to the southwest, going towards the Neillsville area. But, the strangest sighting he described to me, was of a possible 'saucer,' hovering over the field while he was hunting in a neighbor's woods, across from our house! This incident occurred only months after we had moved onto the property. I really wonder if there's a connection to the lights we saw and this 'saucer'!

This next event proved to be the most exciting for me, because I discovered our neighbor, just a few miles east of us, had witnessed the same red lights that I had seen on the 28th of February. Rodney describes the sighting this way: As he was walking back to the house from his garage, he paused for a moment to relieve himself, when something caught his eye at the top of the electrical pole. Thinking that it was a light on the pole, he looked closer, when suddenly, another light lit up, then another. He was now seeing four to six, horizontal, red lights, about the size of baseballs. Within seconds, they began to turn off, one at a time. Hardly believing it, he just stood there, looking into the sky. Then suddenly, the red lights began to appear again, one at a time! But, this time, they were at a forty-five degree angle. Feeling slightly freaked out, he rushed up to the house to get his wife. By the time they both got back outside, the red lights were gone, and all that remained were the 'sparkles'!

Another witness whom I talked with by phone, described to me the very same 'sparkles' above his house. He also lives close to Lindsey. I asked him what he thought they were, and he said, "I don't know what the hell they were, but I know it wasn't the military."

Now I'm sure we're not the only ones that have seen these strange lights near Lindsey, Wisconsin. It wasn't our imagination, and we don't believe it was an aircraft, such as a plane, jet, copter, hot air balloon, glider, flare, etc. None of the lights we observed had sound, and they appeared to be near the tree-top level. The way we look at it, the lights we have seen are 'unidentified.'

So, friends and neighbors, keep checking the skies, and keep an open mind about what may be out there. Then, maybe some day, we can identify the mysterious Lindsey lights.

15

It Was Only a Dream

From the very first sighting of strange lights in 2004, I found myself having difficulty falling asleep. It could have been because my mind refused to let go of the experience, and I kept rehearsing it over and over, in my head. Nonetheless, I was sitting up at all hours of the night by the woodstove. Gene would be snoring throughout the night, and because we had tandem bedrooms, it was a wonder that Amanda could get any sleep.

To help my sleepless hours pass, I would listen to my tiny transistor radio with earphones, changing the stations with a press of a button until I ran across a song I liked. The hours went by slowly, and dawn finally approached. Around five or six in the morning, my eyes would get heavy, and I would collapse onto the couch, and sleep for just a few hours.

I was well aware of Gene getting up between seven and eight, and putting water on to boil in the teakettle for coffee. I tossed around on the couch for a bit, but could not relax enough to fall back to sleep. How many more nights will this go on?

Then, to make matters worse, Amanda got into the habit of watching a movie, and falling asleep on the couch. I tried to get some shut-eye on the loveseat, so that Amanda wouldn't be in the living room by herself. When I noticed that she had fallen asleep, I would get up and turn the TV off. Because Gene would be snoring in the next room, I would have to shake him awake, and insist he roll over and stop snoring.

Of course, I didn't really feel the best about Amanda abandoning her small bedroom to sleep on the couch every night, and she certainly didn't want to give up watching a favorite movie before bed. So, I decided to buy a large air mattress, and lay it on the living room floor. That way, we could both be more comfortable, and maybe get more sleep. Although this might sound a little crazy, our family's new way of sleeping seemed to be working out pretty well.

Then one night, I don't remember exactly when, but it was during the same time we were seeing all those strange lights, that something out of the ordinary occurred.

Gene had said goodnight, and had gone off to bed. Amanda and I put a movie into the DVD player to watch. I laid down on the couch, and Amanda climbed under her blankets on the air mattress beside me. Somehow, I must have suddenly drifted off to sleep, because I quickly began to dream.

In the dream, I was sitting, or laying back, in what resembled a dentist's chair. There were only a few people in the room with me. A couple were sitting on bench-like seats against the walls. Then, someone came up to me, and began inserting something long into my left nostril.

At that time, I began snorting, and gasping for breath, and yelling, "I can't breathe! Take it out! I can't breathe!" Little did I realize, that I was actually yelling, and gasping for breath so violently, that I was screaming in my sleep. All the commotion suddenly shook me awake, with Gene yelling out of the bedroom, "Hey, are you okay out there?"

I looked down at Amanda as she laid there looking up at me, with a startled look on her face. "Mom, are you alright?"

"Yeah, I think so," I answered. "I just had the weirdest dream. I dreamt someone was putting something up my left nostril, and I couldn't breathe!"

As I began to fully come around from the whole incident, I felt very odd. I never had a dream before in my life that shook me awake, and left me gasping for air. And, it all happened so quickly, that I don't remember even falling asleep.

In some of the UFO material I had been reading recently, some of the witnesses of strange lights, described dreaming of sitting or laying, in what resembles a dentist's chair, while being probed in the eyes, ears, or nose. Could I have had the dream because of what I read? Or was it for some other reason?

It was only a dream, but a dream I will most likely never forget.

16

Doctor, Doctor

One night in early December 2003, I received a call from my mother who was crying and saying that my dad had fallen down, and they had called the ambulance to take him to the Marshfield Hospital. Then, I telephoned my sister who lives eighty miles away, by Lake Holcombe. After I explained to her what happened, I quickly got dressed, and met him in the emergency room.

He was a stubborn man in his late eighties, who

disliked doctors, and being in the hospital. After his evaluation, they discovered he had a few cracked ribs. They put him on high doses of pain medication, which caused him to not sleep, and he became delirious. For that reason, my sister, Barb, and I spent a whole week at the hospital, day and night at his side. We slept on chairs in his room.

Because his health condition was deteriorating from kidney failure, due to his diabetes, he couldn't go home, and had to be placed in the Marshfield nursing home. On my way to and from work in town, I would stop to see him.

I remember one day, while my dad was sitting in his wheelchair, I sat upon his bed, and was talking about the birds on his feeder outside the window. We talked about the weather, and the price of gas going up to two dollars a gallon. When we had run out of things to talk about, I brought up the fact that my left arm had been aching for sometime, and that the tingling had traveled down my arm into my hand. I didn't want to worry him, but I told him if it didn't get better, I would have to go to the doctor.

To change the subject somewhat, I decided to bring up the topic of the strange lights Amanda and I had been seeing out by our house. I described the strange 'moon'-like object, that had hovered above the road about a month ago. He didn't say much, he just shook his head and grinned. I think he found it quite entertaining.

In the weeks to come, my sleep became very restless. My arm was aching so bad, I couldn't get comfortable enough to fall asleep. Then, the funny prickling, and tingling, began to travel down my left leg. At that point, I began to get concerned. I looked the symptoms up in my health books I had at home, and thought I was maybe having a heart attack or stroke, from the stress of being worried about my dad.

I made an appointment to see my doctor, at which time they eliminated the possibility of a heart attack or stroke, but did agree that I had a good deal of stress and anxiety, that probably contributed to the pain in my arm.

By early June 2005, the symptoms weren't getting any better. In fact, they appeared to be getting worse. I was now experiencing pain throughout my shoulders, neck, and back, along with headaches and nausea.

I remember the morning of February 29, 2006. It was the morning after I saw the row of six red lights across from our house. Gene had left for a job he was doing on the side at the time, and Amanda was asleep in her bed. We had moved her bed into our bedroom just recently, so she could watch the TV in our room.

I was preparing to take a walk on the trail in the woods with Trixie, and was standing at the stove waiting for the water to boil for coffee, when I felt a sudden rush of extreme nausea. I reached for the teakettle, and shut off the burner. The nausea persisted, and appeared to be getting worse. I became very nervous. I hoped I wasn't about to pass out. I began sweating, and feeling quite ill. "This is not good," I thought out loud.

I started pacing, and then walking around the kitchen table. I tried to clear my mind of all thoughts. What could have suddenly made me feel so sick? As I continued to walk in circles around the table, I began to feel slightly better. But, when I stopped, I began to feel worse again.

Trying not to awaken Amanda, who I could see sleeping through the open door of the bedroom, I had to do something and I had to do it fast. I crossed the living room, and pulled the bedroom door shut.

I decided to grab the tape player, set it on the kitchen floor, and pop in the most relaxing music tape I had. So I would not disturb Amanda, I got down on my knees, and put my head down between my arms, pressing my ear up against the speaker.

I couldn't believe what was happening! I felt so bad! Gene was gone, and I was alone with my twelve-year-old daughter. This feeling needs to go away, and it needs to go away immediately. When the tape came to the end, I swiftly turned it over and pressed "play."

As the soothing music filled my mind, the feeling of nausea started to dissipate. I got up from the floor slowly, feeling weak and shaky and stepped out the door for fresh air. "That's it," I thought. "I need to go back to the doctor. They need to find the reason why I feel this way."

After seeing several doctors in general medicine, a physical therapist, an oral surgeon, to check a tooth for the root canal I recently had done, and a neurologist, there was still no diagnosis made.

I was put on different antidepressants for the pain, and to help me sleep better. My doctor recommended that I see a psychologist, but I declined. I was beginning to experience bad side effects from the antidepressants, so I discontinued them shortly after they were started. I resumed taking Tylenol twice a day.

Tests were run for all types of serious health conditions, and an MRI was preformed to check my neck and spine. Everything came back negative. After more than two years of return visits to the doctors, with my complaints of body pain and nausea, I decided to research the cause on my own.

I took the information I gathered to my neurologist, saying, "I think I could have fibromyalgia." Then, I handed him the literature about the disease.

He looked at me, and said with a chuckle, "You can't diagnose yourself."

"Then, you'll have to do it for me. I have all the symptoms of fibromyalgia. There's nothing left," I replied.

And so it was, in late August of 2006, I was diagnosed with fibromyalgia, for which the cause remains unknown. I can't help but think, that either the stress of taking care of my father in the nursing home, or the sightings of the strange lights, had something to do with my failing health. And, because both events happened at the same time, there is no sure way of knowing. I do believe deep down, that the sightings of some of the 'lights,' at close range, cannot be totally dismissed! For this reason, the mystery of the Lindsey lights will be with me forever.

CONCLUSION

One day, more than twenty years ago, I was watching UFO documentaries on satellite TV, and I came across a program by accident. What they were talking about caught my attention, so I listened quite intently, as the people being interviewed described their encounters with alien beings.

It made me shutter with goose-bumps. Could what they'd been saying really be true? Should I believe in such nonsense? It might be that all these people wanted was attention. At least, I was hoping that's all it was, because if such things were happening to them, they could happen to anyone! The thought sent true fear into my whole being.

It wasn't long after I had watched the UFO programs about people being abducted by aliens, that I saw a report on the world news, that sent fear through me again. This time, it was about cattle being found mutilated in farm fields in Wisconsin and Minnesota. The report said the mutilation might have been done by some occult group for ritual purposes, but other facts at the site of the mutilations raised questions about the real cause.

As a young adult, I remember watching a couple UFO movies with my friends in the theater. You could say that these "true stories" scared the hell out of me! I wanted to put such possibilities out of my mind.

For almost twenty years after that, I was able to distance myself from the alien/UFO topic. Aside from the occasional joke about little green men from Mars, or the science fiction movies, like *E.T. The Extra Terrestrial*, and *Close Encounters of the Third Kind*.

Then, in October of 2004, my young daughter and I witnessed 'unidentified lights' in the sky, though at that time, I did not believe that what we were seeing were UFOs.

Because of the number of these different sightings of

'unidentified lights' that our family witnessed between 2004 and 2007, I had to research the UFO topic. This is what I did. Putting all my past fears of alien abductions to the side, I approached the subject with an open mind. I now had a very good reason to believe that what other people said they saw twenty years ago, could have actually happened.

Now, I just needed to be sure of what it was we had been seeing. So, that is why I began to read as many books as I could about the subject that had instilled so much fear into me years ago. I also contacted a couple paranormal investigators, to no avail. I even went as far as to report our sightings to MUFON (Mutual UFO Network). I guess you could say, I never got a conclusive answer from anyone, which is why I have been drawing my own conclusions.

I got up the courage to write about what our family had seen in the form of this book. I did this not for attention, but to look for an answer.

I do believe that all of God's mysteries will someday be revealed to us, and then, maybe even the mystery of the Lindsey lights will be solved.